Celestial

Love Song

a coloring book with romance,
the sun, the moon, planets, and stars

ORIGINAL ILLUSTRATIONS BY

Laura Medeiros

Tip:
When using markers,
to prevent the color from bleeding through
to the next design, place a blank sheet
under your page.

Celestial Love Song, a coloring book with romance, the sun, the moon, planets, and stars.

Once upon a time in a Sky full of STARS...

© Laura Medeiros

© Laura Medeiros

SOME planets could be TRUSTED to stay on the PATH

© Laura Medeiros

while others NEEDED
Guardians
to guide them on their
WAY

Our Hero, Dashing & Space ..... Traveler and Planet-Herder

Always focused on his task

HE never looked UP

© Laura Medeiros

One day he looked...UP

SEEING THE SUN

for the very first time

©Laura Medeiros

HE fell in Love

His HEART Flew across the sky to her in an Instant

© Laura Medeiros

she was was BEAUTIFUL with dawn's first blush

© Laura Medeiros

...and **FIERCE** as she **SET**

©Laura Medeiros

Moon pursued Sun.
Chasing HER across the sky

© Laura Medeiros

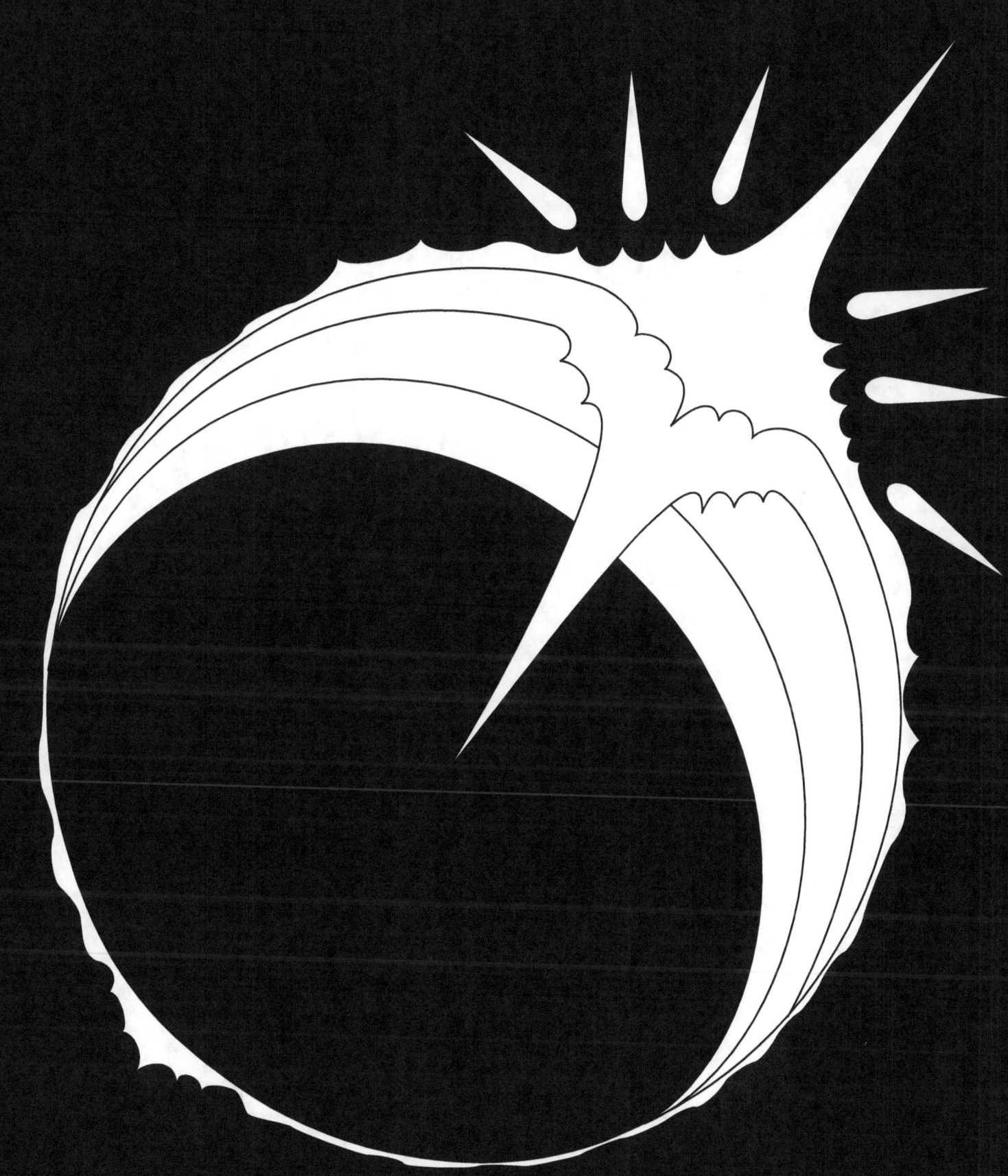

they lived apart in the sky but together in their HEARTS

their

Love

continues

forever

© Laura Medeiros

the sky is full of LOVE stories

Coloring Books

BY

Laura Medeiros

Celestial Love Song

Once upon a time, in a sky full of stars, the moon fell in love with the sun. A love story as old as time told with sixty original drawings and patterns, featuring stars, planets, the moon and the sun. Presented full bleed, printed on one side, with several solid black background images so that your coloring really stands out.

When Octopods Dream

A whimsical collection of octopus and tentacle images for people who like to color.

Thirty original, unique designs and patterns for those who look beyond the ordinary. Printed on one side, and repeated so each image is available to color twice.